Hilbaty Estephany Rodrigues da Silva

Building knowledge with rural young people from Botucatu and the region

Hilbaty Estephany Rodrigues da Silva

Building knowledge with rural young people from Botucatu and the region

Learning about agroecology in terms of its ecoproductive, political-economic and socio-cultural principles

ScienciaScripts

Cover image: www.ingimage.com

This book is a translation from the original published under ISBN 978-3-330-77219-9.

Publisher:
Sciencia Scripts
is a trademark of
Dodo Books Indian Ocean Ltd. and OmniScriptum S.R.L publishing group

120 High Road, East Finchley, London, N2 9ED, United Kingdom
Str. Armeneasca 28/1, office 1, Chisinau MD-2012, Republic of Moldova, Europe
Managing Directors: Ieva Konstantinova, Victoria Ursu
info@omniscriptum.com

Printed at: see last page
ISBN: 978-620-8-64110-8

SUMMARY

1 INTRODUCTION

The earliest use of the word agroecology is related to agroecological zoning, which is the territorial demarcation of the acceptable violation area for a given crop, based on the soil and climate characteristics essential for its development (FEIDEN, 2005).

However, Guzm àn, says that:

Agroecology is becoming fashionable, being used as a mere technique or methodological tool to better understand the functioning and dynamics of agricultural systems and to solve the large number of technical-agronomic problems that conventional agricultural sciences are unable to clarify. However, this restricted dimension - which is gaining a lot of ground in the world of research and teaching as essentially academic knowledge - totally lacks socio-environmental commitments. In this way of understanding agroecology, social variables function to understand the entropic dimension of the deterioration of natural resources in agricultural systems (GUZMAN, 2005, p. 103).

In this context, agroecology also plays a fundamental role not only in training within universities, but also in schools, institutes, research companies, NGOs, technical assistance and rural extension organizations, associations and public agents state and federal agricultural development agencies. However, these sciences and experiences must be aimed at and built for those who are really interested in this knowledge, such as rural youth and rural men.

However, in order to conceive of education from the countryside and for the countryside, it is necessary to mobilize and put into practice ideas and concepts that have long been part of common sense, such as the empirical knowledge built up by family farmers, traditional peoples and communities, and more than that, it is necessary to deconstruct paradigms, prejudices and injustices in order to reverse the historically established learning inequalities between the countryside and the city (BRASIL, 2007).

In view of the above, NGOs have a fundamental role to play in building agroecological knowledge and disseminating agroecology, training agents of change. The role of NGOs is to diagnose the ecological, economic and socio-cultural demands and characteristics of communities and to build agroecological knowledge in a participatory and collective way, based on the endogenous characteristics of that community where the knowledge must be experienced in communities, settlements and all places where an ecologically correct agricultural proposal is desired, helping them to understand the importance of the land, family farming and staying in the countryside.

2 OBJECTIVES OF THE INTERNSHIP

2.1 general objectives

The aim of the internship is to make the university's commitment to extending the knowledge built within it a reality, contributing to the construction of knowledge in the communities and their development, while the student will also be improving their knowledge and getting to know the work of the third sector, experiencing the day-to-day running of a Non-Governmental Organization (NGO).

2.2 Specific objectives

- Performing duties within the institution, helping with activities and building knowledge together with the community;
- Organize, carry out and systematize the activities carried out through the NGO's projects;
- Carrying out charitable tasks in order to raise funds for the IGM's self-sufficiency.

3 THEORETICAL FRAMEWORK

3.1 Agroecology

Agroecology, rather than purely dealing with the ecologically responsible management of natural resources, is established as a field of scientific knowledge which, based on a holistic approach and a systemic approach, aims to help societies redirect the altered course of social and ecological coevolution, in its most diverse interrelationships and mutual influence (CAPORAL, 2009).

However, for Altieri (1998), agroecology provides a methodological framework for a deeper understanding of both the nature of agroecosystems and the principles according to which they function. It is considered to be a science that involves theory and practice, politics and social movement, which goes beyond the boundaries of universities and is based on the ecoproductive, politico-economic and socio-cultural tripod.

It began to be disseminated as a movement opposed to the chemical inputs produced in the Green Revolution, valuing the use of organic matter and other cultural practices that favor biological processes and that benefit the peasant, valuing his culture and traditional knowledge, as well as keeping the farmer in the countryside, strengthening his ties with the land, feeding his family and increasing his income.

This new way of practicing more sustainable agriculture brings with it numerous challenges, such as the environmental, economic, social, territorial and technological ones, all of which are interconnected and can only be achieved through a fair, economically viable and ecologically correct society (ALTIERI, 1998).

Over the years, knowledge about agroecology has been systematized and consolidated, combining traditional and popular empirical knowledge with scientific-academic knowledge, fighting, resisting and advancing gradually but considerably in the awareness of young agents of change.

3.2 Solidarity economy and the collective environment

In recent decades, the solidarity economy has emerged as an alternative for generating work and income for workers exposed to unemployment, underemployment, precarious or informal work, in other words, for those who have been excluded from the capitalist market (OLIVEIRA & ZANIN, 2011).

These are collective production projects, grassroots cooperatives, recyclable materials collection and recycling cooperatives, production, marketing and consumption networks, financial institutes aimed at grassroots solidarity enterprises, self-managed companies, family farming and agro-ecology cooperatives, service cooperatives, among others, which make local economies dynamic, ensure decent work and income for the families involved, as well as promoting environmental care and preservation (FBES, 2006).

According to the Brazilian Forum for Solidarity Economy (FBES), Solidarity Economy (SE) is expressed in organization and awareness of responsible consumption, strengthening relations between countryside and city, between producers and consumers, and enabling more critical and proactive action by consumers on quality of life, food and interest in the direction of development related to economic activity (FBES, 2006).

In this context, collective work experiences emerged, organized in the form of cooperatives, associations, exchange clubs, self-managed companies, cooperation networks, among others (CHAVES & PINTO, 2007).

For Chaves & Pinto (2007), SE is gradually becoming a new alternative for income generation, inclusion and social empowerment as an important response not only for workers but also for poor communities in relation to the transformations that have taken place in the world of work. This new production practice favors collective work, self-management, social justice, care for the environment and responsibility for future generations.

3.3 Rural education, rural youth and the pedagogy of alternation

Questions about the exodus and ageing of the rural population are increasingly gaining repercussions; In addition to the increase in life expectancy in recent decades, another important factor contributing to the aging of the rural population is the exodus of young rural people to the urban environment, where we believe that this movement is revealing in the light of the following situation and the perception that young people can transform the reality in which they live, which is an indicator that we are going through an accelerated period of change in which uncertainty seems to predominate (PUNTEL; PAIVA; RAMOS, 2011).

The recognition that people who live in the countryside have the right to an education that is different from that offered to those who live in the cities is current and modern, recent and innovative, and has gained strength since the National Education Council instituted the Operational Guidelines for Basic Education in Rural Schools, so this recognition goes beyond the notion of geographic space and includes the cultural needs, social rights and integral formation of these individuals (BRASIL, 2007).

In order to meet these specific needs and offer excellent teaching that is suited to the lifestyle, way of thinking and production of the populations identified with the countryside, such as farmers, ranchers, extractivists, fishermen, river dwellers, caiçaras, quilombolas and rubber tappers, a new concept of education has emerged: Countryside Education (BRASIL, 2007).

However, within Countryside Education is the Pedagogy of Alternation, which is a specific methodology that works with young rural people and has successful cases of innovation in the field of agroecological production, family and organic farming, almost always developed on the basis of the precepts of an economy that wants to be more solidary and cooperative (OLIVEIRA, 2012).

The Pedagogy of Alternation consisted, in its European origins, in the 1930s, of a methodology for organizing school education that sought to combine different educational experiences, distributed over different times and spaces, with the aim of providing a more suitable education for those who live on the land and from the land, arriving in Brazil in the mid-1960s, under the perspicacious leadership of an Italian priest who was looking for a teaching method for the school he had been called to direct, in the south of Espirito Santo. (OLIVEIRA, 2012, p. 05).

3.4 The role of NGOs in Brazil and in agroecology

In Brazil specifically, NGOs are part of the new social movements and have tended towards human development and broadening citizen participation. They have a wide range of themes, ranging from environmental organizations and feminist groups to child and adolescent protection organizations (BUARQUE & VAINSENCHER, 2001).

The history of the formation of ecological movements in Brazil, which culminated in what we know today as Agroecology, is a systematized science and a recent social movement, but as a practice it is thousands of years old, converging with the worldwide strengthening of the struggle of non-governmental organizations (NGOs) materialized in their participation in international forums for discussion on the environment and development since the 1970s (BENTHIEN, 2007).

From the outset, NGO public action gained visibility and transparency by supporting community movements and associations in the most diverse areas of action (defense of human rights, social rights, overcoming social inequality; training/education of agents of social transformation and provision of services to the community, environmentalism) (NEDER, 1998).

Brazilian NGOs play an important role in advising on the collective construction of agroecological knowledge and practices. The establishment and validation of agroecological practices must be a collectively built process, with family farmers, traditional peoples and communities playing a leading role, mainly through the formation of rural advisory and monitoring groups that provide farmers with information on the best way to treat the land, as well as helping them to dispose of their produce and avoid losses (BENTHIEN, 2007).

4 SITE CHARACTERIZATION

The internship was held at the Instituto Giramundo Mutuando (IGM) from June to August 2016, located at Rua Cardoso de Almeida, 1207, in the city center of Botucatu, a Brazilian municipality in the state of Sao Paulo, belonging to the south-central region. It is located at 22°53'09" south and latitude 48°26'42" west . longitudeThe city has a mild climate with average temperatures of 22° C, classified as humid subtropical, and an altitude ranging from 756m to 920m.

The Giramundo Institute, where the work was carried out, is a non-profit, public interest civil society organization that promotes agroecological agricultural production practices and a focus on Sustainable Local Development, always striving to build knowledge with family and ecological farming communities, encouraging sustainable management, respect and preservation of nature, so that disadvantaged communities can find more income and more food through agroecology. Since 1998, the Giramundo Mutuando Institute has been strengthening the agroecological movement for environmentally sound and socially just agriculture, offering technical support for artisanal production, family, community and organic agricultural production, as well as supporting the marketing of these products.

Giramundo's objectives are to carry out study projects, research and courses in general, linked to Sustainable Development, Agroecology and Participatory Methodologies; to hold forums to discuss Public Policies; to bring professionals, organizations, projects and investments in the area of Sustainable Development closer to those interested in the communities; to produce and disseminate technical, scientific and popular information and knowledge.

4.1 History and activities

The Giramundo Mutuando Institute (IGM) began its work in 1998, through the Planeta Limpo Project for municipal selective waste collection and the organization of a waste pickers' cooperative, in partnership with the Cerqueirense Association of the Vital Age - ACERVI, in the municipality of Cerqueira César.

In 2000 it offered organizational and technical-productive support to organic producers in Botucatu and the region, helping with participatory planning for the production and marketing of organic products.

In 2001, it held the 1st International Meeting on Agroecology and Sustainable Rural Development and the Regional Fair of Organic Products, Crafts and Culture of Botucatu and Region. The meeting brought together 600 participants from 14 Brazilian states and 5 countries, with national and international repercussions. The Fair was visited by 1500 people from the region.

In January 2002, the organization held the event "Let's Make the World Everyone's Land", which included various cultural activities, raising awareness among the local population about the issue of sustainable use of Brazilian and world territory.

Also in 2002, in partnership with the Tourism and Environment Secretariats of Botucatu City Hall, the Institute held the Ecological Tourism Generator Workshop, aimed at thinking about tourism in Botucatu through a participatory approach.

In 2003, together with the Avaré city council, he developed a project to promote community gardens in outlying neighborhoods.

Also in the agricultural field, the Giramundo Mutuando Institute has been developing, since July 2001, participatory work to form the identity of the Agroecological Rural Extension professional with students of Agricultural Sciences at the Universidade Estadual Paulista - UNESP in Botucatu, among other universities.

Since 2004, the Institute has been working on the development of the Agroecological Rural Extension Program for Botucatu and the Region - PROGERA. The main objective of this program is to develop and empower family farming in the Botucatu region, based on the principles of Agroecology and Participatory Methodologies, creating conditions that favor the environmental recovery of degraded areas and the Agroecological Transition in small properties in the region. PROGERA also carries out actions in rural settlements in the municipalities of Itapeva and Iaras and *has already* carried out various technical assistance and rural extension activities and training events. These include the 1st and 2nd Local Sustainable Development Forum. However, in 2005, Giramundo held the 2nd International Meeting on Agroecology and Rural Development, which was attended by 400 farmers from all over the state of São Paulo and 600 professionals and students from the field of Agricultural Sciences from all over the country.

In 2007, with the support of WWF-Brazil, it also developed the Gigante Guarani Project, which aimed to recover APPs in the vicinity of the Alambari River, an area of extreme interest for conservation, not least because it is a recharge area for the Guarani Aquifer System. In this project, he mapped forest fragments of great interest for the identification of matrices for the production of forest seeds. In 2010, FEHIDRO approved new funding to continue the work of the Gigante Guarani Project. With the support of UNESP's Biosciences Institute and CNPq, it has contributed to improving the curricula of the Agricultural Sciences, characterizing six higher education courses in Agroecology. Since 2009, it has been carrying out training activities with young people from rural areas with the support of the Municipal Council for the Rights of Children and Adolescents.

In the context of PROGERA, from 2010 to 2011 the identification of demonstration units of Agroforestry Systems was carried out in 11 regional offices of the Articulaçao Paulista de Agroecologia with the support of the Instituto de Economia Agricola de SP (IEA), CNPq and UNESP. From 2012 to 2013, she contributed to the consolidation of family farming demonstration units in: urban agriculture, settlements, women's groups, rural tourism, consumer networks and participatory group certification.

The Institute participates in local, regional and national political bodies such as the Local Sustainable Development Forum, the Regional Food Security Council, the Municipal Environment Council, the Rio Pardo Consortium for Studies and Sustainable Development, the National Agroecology Articulation, the São Paulo Agroecology Articulation, the Brazilian Agroecology Association and the Municipal Council for the Rights of Children and Adolescents (CMDCA).

5 ACTIVITIES CARRIED OUT

5.1 Project: "Youth, Agroecology and Environmental Education: awakening young people to the importance of nature and ecologically-based food production".

5.1.1 Project Summary

This project has been running since 2009 with young people aged between 14 and 18 from rural areas, with the support of the Municipal Council for the Rights of Children and Adolescents (CMDCA).

The activities are carried out weekly, on Tuesdays and Thursdays in the afternoons, and last for 12 months. Central themes are covered, such as agroecology, the production of healthy food, respect for the land, the importance of nature and conscious cooking, all with a critical approach and popular language so that young people are involved, providing conscious training and the collective construction of knowledge.

The project recruits teenagers living in peripheral and rural areas of Botucatu and the surrounding region. The project team disseminates information mainly in public schools located in neighborhoods in need of social assistance. It has the capacity to attend up to 30 young people per training period. The young people must have an interest in the subject, belong to low-income families and be in a situation of social vulnerability, including obvious social problems such as precarious housing conditions, food insecurity, among others related to socio-economic conditions, such as violence and teenage pregnancy.

The teaching methodology used is based on educommunication, which prioritizes the participation and integration of young people, from the choice of topics of greatest interest to the format of each workshop and the materials generated. In the activities, young people are invited to reflect on issues related to agriculture, food production, the impact of development on the environment and health, professions and interpersonal relationships.

Giramundo has a community vegetable garden, which was designed and carried out together with the young people taking part in the project. In addition to fruit trees, there are also some agricultural species, non-conventional food plants (PANCs), medicinal plants and spices. Management is preferably carried out during the project's activities and workshops, bringing young people closer to the principles of agroecology and contact with nature. As well as producing food for the afternoon snack, it also produces fresh spices, such as basil and oregano, to serve the pizzeria - which owns the backyard where the vegetable garden, orchard, nursery and kitchen are located as spaces for participatory and collective experience and construction of agroecological knowledge and practices, environmental education, solidarity economy, etc. using the methodology of Educommunication, Emancipatory Education and Participatory Action Research (IAP), the basic methodologies of research in Agroecology.

Figure 1: Pâmela watering the IGM vegetable garden, Botucatu-SP, 2016.

Figure 2: Sweet potato harvested in Giramundo's backyard during the youth project, Botucatu-SP, 2016.

5.1.2 Activities carried out

The project is currently being developed by a forestry technician, currently studying pedagogy, who acts as a facilitator and educator with extensive experience in agroecology.

At the beginning of the internship, the activities continued to be developed normally on Tuesday and Thursday afternoons with a variety of themes, such as: urban and alternative horticulture practices, debates on films and documentaries, group dynamics, all based on the collective construction of the activities, always valuing the opinions of the young people, in order to sensitize them, developing awareness and intellect for the issues that form the tripod of agroecology, within the ecoproductive, sociocultural and political-economic dimensions.

Figure 3: Agroecology game dynamics carried out with young people in Giramundo's backyard, Botucatu-SP, 2016.

During the course of the internship, the young people were always told about the difference between agroecological agriculture and conventional agriculture, showing them not only in conversations, dynamics, but also in practical activities the best way to farm, respecting life and nature. This takes into account man's relationship with the land, the conscious use of the soil, the diversification and production of vegetables, the balance of fauna and flora and the marketing of agricultural products in both its forms, making it clear which is the best model for production and life.

The workshops held in Giramundo's backyard are always accompanied by a theoretical and dynamic approach to the practice to be carried out, involving agroecological cultivation techniques, fertilization, types of management, identification of plants in the backyard, collection and storage of seeds, soil characteristics, with the help of the facilitators, with the aim of instigating and promoting debates on subjects such as: conscious consumption, healthy eating, acquiring products from agroecology and family farming, encouraging changes in habits in relation to the environment and food.

The trainee's job is to plan, accompany and assist the Forestry Technician in his activities with the young people, giving them support in carrying out their work. The trainee is free to do whatever he or she wants, from helping with workshops and talks with the young people to preparing the afternoon snack, all of which is done with their participation, always prioritizing and providing healthy food and a sustainable awareness.

Figure 4: Preparation of the afternoon snack, always based on a healthy diet, Botucatu-SP, 2016.

Figure 5: Afternoon snack, IGM - Botucatu-SP, 2016.

5.2 Project: "Peasant Youth and Agroecological Transition: protagonism, participation and social movement in building the future."

5.2.1 Project Summary

The project, built on a process of broad articulation with grassroots organizations and social movements, corresponds to the wishes of the organizations of the Fazenda Pirituba II and Zumbi dos Palmares settlements, in the southwest and central regions of São Paulo. It also represents the demands that exist throughout the state of São Paulo, fundamentally from the rural youth movements present there.

This project promotes the training of young people based on the construction of knowledge and the strengthening of Agroecology experiences in Agrarian Reform settlements, carried out through the Pedagogy of Alternation and Freirean presuppositions; Fifty young peasants linked to the MST schools "Institute Laudenor de Souza (Itapeva-SP) and Escola Estadual Popular Rosa de Luxemburg" (Agudos-SP), which include, in addition to a continuous course in Agroecology, thematic workshops, participatory rural diagnosis activities, field days, local, regional and state exchanges, workshops, systematization of experiences, production of teaching materials and pocket videos, as well as events and other activities (CARMO; MOREIRA; STAMATO; 2014).

This partnership will take place through the IGM, the National Institute for Colonization and Agrarian Reform (INCRA), the National Council for Scientific and Technological Development (CNPq), the Landless Rural Workers' Movement (MST), São Paulo State University - Botucatu campus (UNESP) and the Ministry of Agrarian Development (MDA).

The activities are carried out by IGM educators and facilitators, external guests and volunteers and activists from the Landless Movement (MST). The entire schedule for the activities is defined in agreement with the local MST coordinators, so that it is an activity that benefits both sides, taking into account the interests of those involved and, above all, continually seeking to build knowledge collectively and horizontally.

The project also promotes intense statewide exchanges between the young beneficiaries and young people linked to concrete experiences of agroecology in all regions of the state of São Paulo, promoting an intense flow of communication during the training process and its articulation with all the activities of building agroecological knowledge, in order to empower the peasant youth of the settlements worked on, towards their leading role in building the agroecological transition in their areas of life and social organization.

The process of linking the grassroots organizations in these settlements and the respective rural youth movements, with the university and other project partners, will result in the strengthening of education in the Agrarian Reform areas, stimulating educational projects with methodologies geared to the specific characteristics of the countryside, and contributing to the promotion of sustainable rural development and the reversal of the exodus of rural youth.

5.2.2 Activities carried out

The trainee's role at the IGM consists of helping and contributing to the construction of knowledge and activities, as well as having an active voice in collective decision-making and participating assiduously, collaborating with whatever is necessary for the development of a job that is well done, improved, enriching and beautiful.

During the course of the internship, visits were made to each of the rural settlements served by the project over the course of the months, both during community time, which takes place on weekends, and during school time, which takes place when the students are on vacation.

The program for the activities carried out is always built collectively and agreed upon, together with the young people from the project and the MST leaders. Despite their differences, in both settlements the main themes to be addressed by the Giramundo team were the recovery of activities related to the current community time and a workshop on the construction of the course completion work (TCC). The material produced by the students on the agro-ecological transition experienced on their own plot will result in the expansion of agro-ecological science and the knowledge acquired by the students during the course.

Figure 6: Visit to Caio's plot showing the agroecological transition process, Iaras -SP, 2016

The project is scheduled to end at the end of November with the presentation of the Course Conclusion Papers (TCC) and the Regional Agroecology Meeting, where the young people will present their TCCs, receive their certificates of completion at a ceremony and take part in the agroecological fair that will take place at the event. The tasks carried out so far in this module have been: family/plot timeline; drawing of the plot and the systems present, agroecosystem analysis, flowchart of inputs and outputs, income and work, agroecological transition, consumption of non-conventional food plants (PANCS) and identification of spontaneous plants. The compilation of these activities are the themes for the preparation of the CBT.

Figure 7: Recognition of PANCs in the SAF of the Rosa Luxemburgo school, Iaras-SP, 2016.

As an intern, I helped build activities for the penultimate module of the Rosa Luxemburgo settlement project in Iaras-SP, where I facilitated a space on agroecological transition with the help of my fellow students; We visited the plot of a student on the course, where he showed how his transition process was developing, we recognized the PANC's in the school's SAF, we also made a carousel to discuss the agroecological dimensions, we talked about healthy eating and the issue of food waste, as well as how the agri-food system works, and then we put what we had seen in class into practice in the kitchen, preparing food for the June party as the course's cultural evening, and I helped the young people to construct their CBTs,

Figure 8: Start of the day's activities, always with a dynamic, Rosa Luxemburogo School, Iaras-SP, 2016.

Figure 9: Class on agroecological transition, at the Rosa Luxemburogo School, Iaras-SP, 2016.

Figure 10: Lesson on healthy eating and the agri-food system

Figura 11: Group dynamics on agroecological dimensions - presentation by students at the Rosa Luxemburogo School, Iaras-SP, 2016.

Figura 12: Decorations for the June party as a cultural evening at the Rosa Luxemburogo School, Iaras-SP, 2016.

6 CORRELATION WITH THE COURSE

The IGM, the institute where the internship was carried out, works in the field of agroecology, which correlates with the agronomy course not only because of the agroecology subject, but also because of the search for a more sustainable, fair and healthy agriculture. The most important content seen in the classroom for the development of the activities and the awakening of awareness were: The absurdity of modern agriculture and the evolution of production systems, which were extremely important for carrying out the work.

7 FINAL CONSIDERATIONS

The IGM has been working with agroecology for almost 20 years, and as its name implies, it rotates and mutates, sharing and participating in countless stories, favoring exchanges of experience, promoting events, training, courses and working on transformative projects.

Giramundo welcomes interns and trainees where it presents the importance of working in the third sector, with its difficulties, challenges and projects, but also with the success of a long, arduous journey, of struggle and intensified by the persistence to keep alive a link that goes beyond healthy agriculture, but which mainly promotes the well-being of the peasant people, The main area of activity of the Giramundo Institute is Agroecology, which over the years has been highlighted as a science, promoting research and empowering students to prove that this is the best way to use the land. The city of Botucatu, with its peaceful green landscape, where the Giramundo Mutuando Institute is located, nevertheless has some conventional monocultures such as sugar cane and eucalyptus for the production of paper and cellulose.

Brazil is the country with the highest use of pesticides in the world and the city of Botucatu is not left out of this terrible statistic, with its use of chemicals, inputs and fertilizers. However, on the other hand, the city has a large number of organizations willing to reverse this situation, working on research linked to agroecology, with an emphasis on family, organic and biodynamic agriculture, with stores and restaurants that always value healthy food and conscious consumption.

Therefore, the internship was of the utmost importance for me to carry out a job where I was able to have unique experiences, putting into practice the knowledge acquired at university and in life-long experiences, adding knowledge to my professional growth, but mainly to my personal development, exceeding my expectations and allowing me to improve the techniques learned as well as perfecting my curriculum in the area of agroecology, which involves family farming, organic farming, healthy eating and conscious consumption.

In this way, the compulsory internship met its objectives and contributed to my professional development.

Figure 13: Educators and students united to strengthen agroecology and combat the exodus of young rural people

8 THANKS

First of all, I would like to thank God for giving me the health and strength to overcome all the difficulties, allowing me to live this moment, bringing joy to my parents and to everyone who contributed to the realization of this work.

The Universidade Federal Rural do Semiàrido and all its teaching staff, as well as the management and administration, who carry out their work with so much love and dedication, working tirelessly so that we, the students, can count on an extremely high-quality education.

To the Giramundo Mutuando Institute, which gave me the opportunity to do excellent work together, in a harmonious environment, with qualified professionals who were willing to exchange experiences and pass on their knowledge.

I would like to thank my advisor, Vania Christina Nascimento Porto, for her guidance, patience, dedication and teachings that enabled me to carry out this work.

My special thanks go to my family for their love, affection, patience and for sparing no effort so that I could carry on with my studies. As well as my friends, for trusting me and being by my side at all times.

REFERENCES

ALTIERI, MIGUEL. **Agroecology: the productive dynamics of sustainable agriculture** /MIGUEL ALTIERI. - 4.ed. - Porto Alegre : Editora da UFRGS, 2004.

BENTHIEN, P.F. **NGOs and agroecology in Brazil. Abstracts of the II Brazilian Congress of Agroecology**. Rev. Bras. Agroecologia, v.2, n.1, feb. 2007.

BUARQUE, C; VAINSENCHER, S. A. **NGOs in Brazil and the gender issue**, 2001.

CAPORAL, F. R. **Agroecology: a new science to support the transition to more sustainable agriculture**. Brasilia, DF, 2009. p.30.

CAPORAL, F. R; AZEVED, E. O. **Principios e perspectivas da agroecologia.** Instituto federal de educaçao, ciência e tecnologia do paranà -educaçao a distância, 2009.

CARMO, M. S. do; MOREIRA, R. M; STAMATO, B. **Peasant youth and agroecological transition: protagonism, participation and social movement in building the future**. 2014.

CHAVES, D. F; PINTO, I. M. J. **Solidarity economy as an alternative for regional development.** T&C Amazônia, Year V, Number 10, February 2007.

Rural Education: differences changing paradigms. Ministry of Education, Brasilia, DF, 2007.

FEIDEN, A. **Agroecology: principles and techniques for sustainable organic agriculture.** Embrapa Informaçao Tecnològica, Brasilia, DF, 2005. Agroecology: introduction and concepts. Chap. 2, p.53.

Brazilian Forum on Solidarity Economy, 2006.

GUZMAN, E. S. **Agroecology: principles and techniques for sustainable organic agriculture**. Embrapa Informaçao Tecnològica, Brasilia, DF, 2005. Agroecology and sustainable rural development. Chap. 4, p.103.

NEDER, R. T. **As ONGs na reconstruçâo da sociedade civil no Brasil.**

State Reform Society, Sao Paulo, March 26-28, 1998.

OLIVEIRA, M. C. S. B; ZANIN, M. **Solidarity Economy: an evolving theme in Brazilian dissertations and theses**. Revista Brasileira de Ciência, Tecnologia e Sociedade, v.2, n.1, p.181-193, jan/jun 2011.

OLIVEIRA, M. M. **Jovens e Educaçâo no Campo: por uma política de reconhecimento dos movimentos sociais campesinos.** Interview conducted by professor Elionaldo Fernandes Juliao in August 2012 especially for this publication. p. 1-10.

PUNTEL, J. A; PAIVA, C. A. N; RAMOS, M. P. **Situação e perspectivas dos jovens rurais no campo.** Proceedings of the 1st circuit of academic debates. CODE, 2011.

MIX
Papier aus verantwortungsvollen Quellen
Paper from responsible sources
FSC® C105338

Printed by Books on Demand GmbH, Norderstedt / Germany